YOUR KNOWLEDGE HAS VALUE

- We will publish your bachelor's and
  master's thesis, essays and papers

- Your own eBook and book -
  sold worldwide in all relevant shops

- Earn money with each sale

Upload your text at www.GRIN.com
and publish for free

Imprint:

Copyright © 2016 GRIN Verlag, Open Publishing GmbH
Print and binding: Books on Demand GmbH, Norderstedt Germany
ISBN: 9783668366244

This book at GRIN:

http://www.grin.com/en/e-book/347113/nesting-behavior-and-habitats-of-the-
stingless-bee-trigona-iridipennis

Prem Jose Vazhacharickal, Sajan Jose K.

Nesting behavior and habitats of the stingless bee Trigona iridipennis Smith in Kerala

The current status of knowledge

GRIN Publishing

Nesting behaviour and habitats of stingless bees (*Trigona iridipennis* Smith) in Kerala: current status of knowledge

ACKNOWLEDGEMENT

Firstly we thank **God Almighty** whose blessing were always with us and helped us to complete this research work successfully.

The first author is extremely grateful to **Dr. Sajeshkumar N.K** (Head of the Department, Biotechnology) for the valuable suggestions, support and encouragements.

The first author would wish to thank beloved Manager **Rev. Fr. Dr. George Njarakunnel,** respected Principal **Dr. V.J. Joseph,** Bursar **Shaji Augustine,** Vice Principal **Fr. Joseph Allencheril**, and the Management for providing all the necessary facilities in carrying out the study.

We lovingly and gratefully indebted to our teachers, parents, siblings and friends who were there always for helping us in this project.

Prem Jose Vazhacharickal* and Sajan Jose K

*Address for correspondence
Assistant Professor
Department of Biotechnology
Mar Augusthinsoe College
Ramapuram-686576
Kerala, India
premjosev@gmail.com

Table of contents

Table of tables

List of abbreviations

PVC	: Polyvinyl chloride
SPSS	: Statistical package for social sciences

Nesting behavior and habitats of stingless bees (*Trigona iridipennis* Smith) in Kerala: current status of knowledge

Prem Jose Vazhacharickal[1]* and Sajan Jose K[2]

* premjosev@gmail.com

[1]Department of Biotechnology, Mar Augusthinose College, Ramapuram, Kerala, India-686576

[2]Department of Zoology, St. Joseph's College, Moolamattom, Kerala, India-685591

Abstract

Stingless bees are highly social insects which populated the tropical earth 65 million years ago longer than honey bees. They are limited to tropics and subtropics lacking venom apparatus and cannot sting. Impacts of anthropogenic influences on honey bees were already reported. Recent studies also showed that the nesting behaviour of *Trigona iridipennis* Smith in natural habitat also vary due to interaction, pheromones and environmental stimulus. A little is reported so far about the various natural and domesticated nesting of the *Trigona iridipennis* Smith in Kerala. Based on these back ground, our objectives of this study were to 1) to characterize the Meliponiculture 2) to identify the various natural habitats and domestication materials for nest construction and different types of nests used across Kerala. Various beekeeping methods preferred by farmers across Kerala for the cultivation of *Trigona iridipennis* Smith. Each nest has its own advantage and disadvantage. During the survey, the most preferred one's were wooden box. Even then according to the easy availability and production cost different nests like earthen pot, bamboo nodes, coconut shell, PVC pipes etc were used. The most preferred natural nesting sites by *Trigona iridipennis* Smith were mud/stone walls in our study. Being a social insect, *Trigona iridipennis* Smith shows great diversity in nesting pattern in natural habitats as well as anthropogenic habitats. The shift towards anthropogenic habitats than natural ones may be due to destruction of natural habitats as well as the availability of manmade habitats.

Keywords: *Trigona iridipennis* Smith; Anthropogenic habitats; Entrance tube; Meliponiculture.

1. Introduction

Stingless bees are highly social (eusocial) insects which populated the tropical earth 65 million years ago longer than honey bees (Camargo and Pedro, 1992). They are limited to tropics and subtropics lacking venom apparatus and cannot sting. They have well developed mandibles using which they can bite and keep away intruders. Vestigial sting, presence of penicillum (a bunch of stiff setae on the outer apical margin of hind tibia), reduction and weakness of wing venation are the three major characters of stingless bees. The stingless bee found in Kerala is *Trigona iridipennis* Smith also called 'dammer bees' as they collect a kind of resin from plants to construct their nest. They are locally known as 'Cherutheneecha' in Malayalam has also called mosquito bees and are found all over India (Singh, 2013). The name *Trigona* refers to their triangular abdomen and '*iridipennis*' refers to their iridescent wings. Stingless bees are small to medium sized with vestigial stings, shows social level of organization and pollinators of flowering plants in the tropics (Ramanujam et al., 1993; Heard, 1999; Amano et al., 2000; Raju et al., 2009). The *Trigona iridipennis* Smith is considered as highly eusocial bees living as perennial colonies. They belong to the family Apidae and subfamily Meliponinae capable of domesticating in hives can moved from places with non-stinging characteristics (Amano et al., 2000; Franck et al., 2004, Rasmussen, 2013).

Keeping of stingless bees is called Meliponiculture and *Trigona iridipennis* Smith are kept in India for centuries for the high medicinal value of honey as well as propolis and bee wax (Cortopassi-Laurino et al., 2006; Choudhari et al., 2012; Kumar et al., 2012; Andualem, 2013; Choudhari et al., 2013; Rasmussen, 2013; Virkar et al., 2014). The average of the honey ranges from 400-600 g per colony and fetch around 2000 INR (12 US$) per kilogram due to its less productivity as well as high demand from pharmaceutical sector (Kumar et al., 2012). Honey of the stingless bee, *Trigona* spp., is usually a highly praised apitherapeutic agent used as a panacea against dozens of ailments in Ethiopia. Among the most common uses of stingless bee honey are to treat stomach disturbance, cough, tonsillitis, sore throat, stomach and intestinal ulcers, cold, disease of the mouth, mucus membrane, and as a wound dressing due to its antimicrobial activity (Garedew, 2004; DeMera and Angert, 2004; Boorn et al., 2010).

Impact of anthropogenic influences on honey bees were already reported by Basavarajappa (2010). Recent studies also showed that the nesting behaviour of *Trigona iridipennis* Smith in natural habitat also vary due to interaction, pheromones and environmental stimulus (Basavarajappa, 2010; Pavithra et al., 2012). A little is reported so far about the various natural and domesticated nesting of the *Trigona iridipennis* Smith in Kerala.

According to Singh (2013) and Mohan and Devanesan (1999), the nest of *Trigona iridipennis* Smith consist of entrance, brood, food storage pots, resin dump, waste dump, pillars, connectives, involucrum, and batumen. Difference in the entrance depending on the species (Chinh et al., 2005; Bänziger et al., 2011; Pavithra et al., 2012; Danaraddi et al., 2012) as well as tree preference for building nest (Marisa and Salni, 2012). Worker bees secrete wax which are mixed with resin collected from various trees especially red wood, jackfruit, bread fruit forming a dark coloured sticky substance called cerumen (Nair and Nair, 2001; Nair, 2003; Marisa and Salni, 2012; Rasmussen, 2013; Singh, 2013). Usually the brood is in the form of cluster which contains larvae. Honey and pollen pots are larger than brood cells. The brood cells are oval in shape, arranged in clusters connected to each other by short pillars and connectives. The worker bees are a bit darker in colour with a mean body length of 4.07 mm. The queen have golden brown colour with a mean body length of 10.07 mm (Devanesan et al., 2009). Based on these back ground, our objectives of this study were to 1) to characterize the Meliponiculture 2) to identify the various natural habitats and domestication materials for nest construction and different types of nests used across Kerala.

2. Hypothesis

The current research work is based on the following hypothesis

1) The nesting behaviour and habitats of stingless bees in Kerala may vary across different locations
2) Farmers in Kerala grow stingless bees in conventional and non-conventional methods.

3. Materials and Methods

3.1 Study area

Kerala state covers an area of 38,863 km^2 with a population density of 859 per km^2 and spread across 14 districts. The climate is characterized by tropical wet and dry with average annual rainfall amounts to 2,817 ± 406 mm and mean annual temperature is 26.8°C (averages from 1871-2005; Krishnakumar et al., 2009). Maximum rainfall occurs from June to September mainly due to South West Monsoon and temperatures are highest in May and November (Figure 1).

3.2 Study design and data collection

Twenty five stingless bee samples were collected from twenty one locations in twelve districts across Kerala using information's collected from various agricultural departments, farmers and beneficiaries (Figure 2). Few nests on wall and tree trunk were made exposed to learn the internal nest architecture. Hundred and twenty farmers engaged in Meliponiculture were interviewed and observed. A questionnaire was prepared and distributed among the farmers to learn their depth in Meliponiculture methods and techniques and to study the current status of Meliponiculture in Kerala. The plus and minus points of various Meliponiculture techniques were also mentioned.

3.3 Statistical analysis

The survey results were analyzed and descriptive statistics were done using SPSS 12.0 (SPSS Inc., an IBM Company, Chicago, USA) and graphs were generated using Sigma Plot 7 (Systat Software Inc., Chicago, USA).

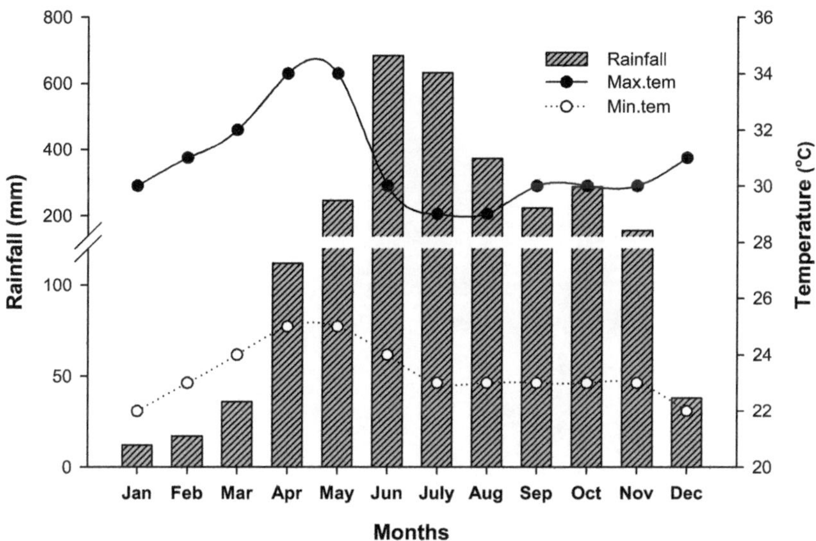

Figure 1. Mean monthly rainfall (mm), maximum and minimum temperatures (°C) in Kerala, India (1871-2005; Krishnakumar et al., 2009). Author's own work.

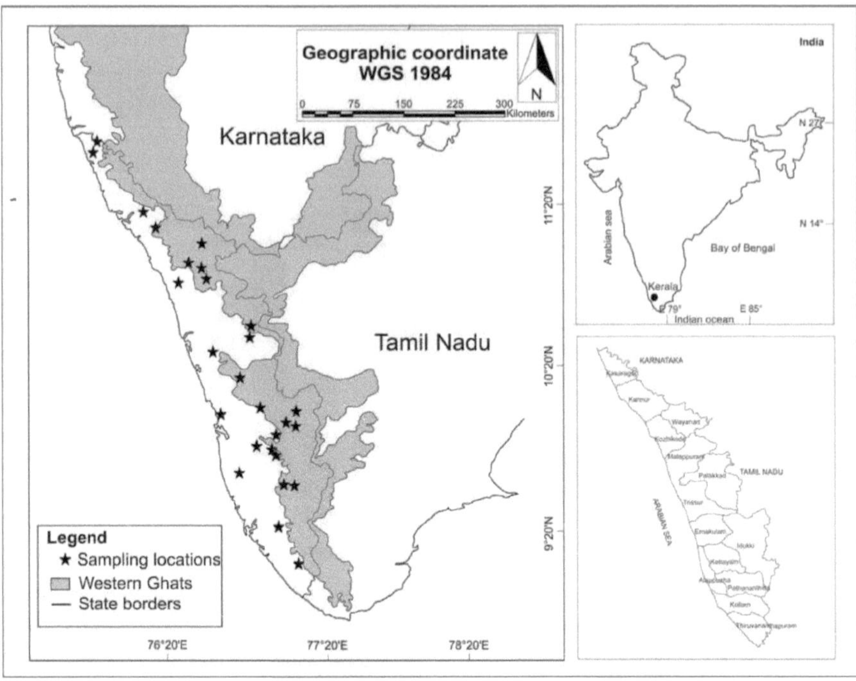

Figure 2. Map of Kerala showing various sample collection points marked as a star symbol during the study (Authors own image).

Figure 3. Stingless bee a) worker foraging on euphorbia flower; b) drone (top) and worker (bottom); c) brood with queen cell; d) entrance tube with fresh resin at the tip; e) food pots (honey) and pollen pots intermingled (Authors own image).

Figure 4. Different types of bee nests a) hives placed on the stand; b) PVC pipe nest; c) earthen pot nest; d) nest in coconut shell; e) bamboo pole nest; f) earthen bowl nest exposed; g) log nest; h) nest in garden pot; i) two tier earthen bowl nest (Authors own image).

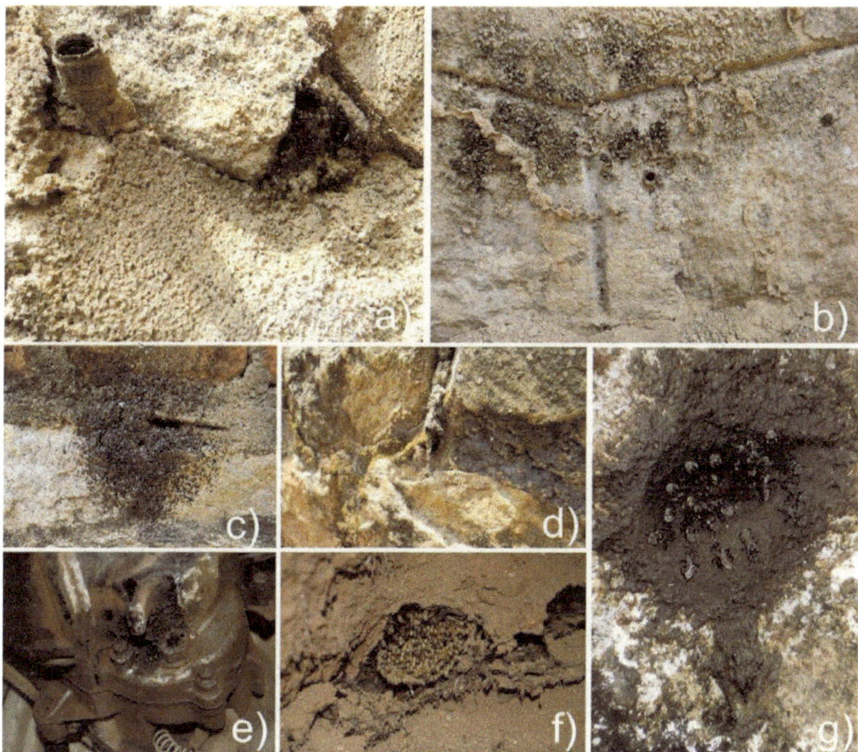

Figure 5. Various types of nests and nest entrance among stingless bees a) upward directional entrance tube; b), c) and d) entrance tube on stone walls; e) entrance tube on scooter engine ; f) exposed mud wall nest; g) guard bees near the entrance tube (Authors own image).

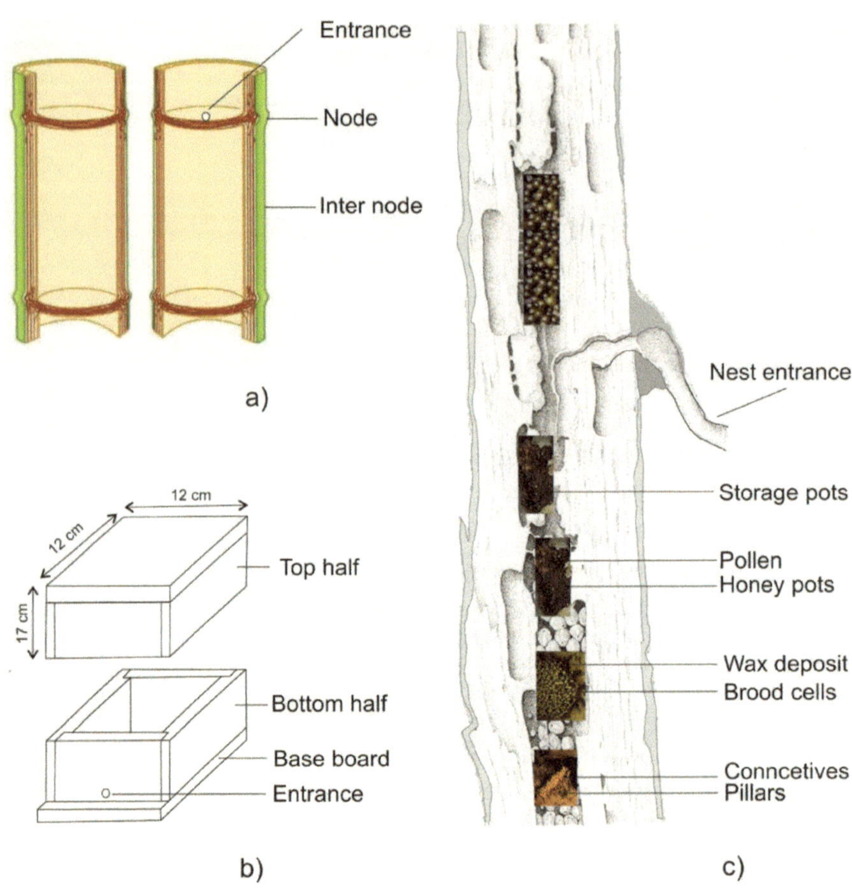

Figure 6. Schematic representation of stingless bees in differ habitats a) bamboo pole; b) wooden box; c) inside a tree cavity.

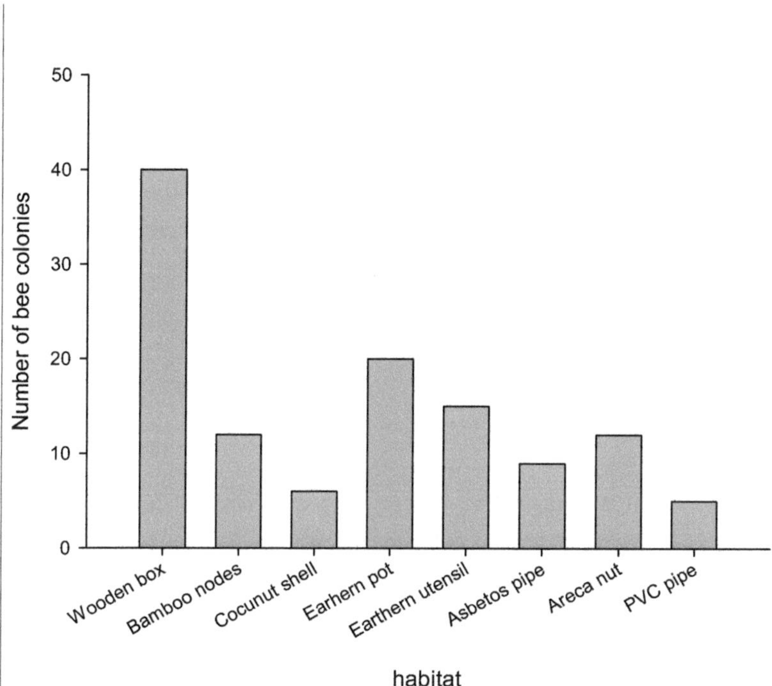

Figure 7. Preference of farmers (n=15) for the cultivation of stingless bees in Kerala using different materials. Authors own work.

4. Results

4.1 Nest architecture and characteristics

The nest is the central place from which stingless bees mate, forage and pass through life stages. Nests are immobile fixtures and potentially long live (Michener, 1974; Roubik, 1989). Wall cavities and tree cavities are the major nesting sites for *Trigona iridipennis* Smith. They also make their nests in switch boards, telephone posts, discarded pipes etc. Though rare they were also found construct their nest on lorry cabin and also in scooter engine box. One colony even made their nest in a rain coat which was hanged in the car-porch in the off season. They are highly anthropophilous and more active in vicinity of human surroundings. A conspicuous entrance tube is found in front of the nest (Figure 3).

4.1.1 Entrance tube

The stingless bee nest is always characterized by a nest cavity, typically provided with a very narrow opening facilitating defence (Kolmes and Sommeijer, 1992). The simplest stingless bee nest entrance protrudes slightly from the base of the entrance hole. Nest entrance is related to defence and foraging (Biesmeijer et al., 2005). It is made up of cerumen, a mixture of wax and resin. In addition to wax and resin foreign materials like grease, smoke particles, fibres, mud, dust particles etc may be seen on the entrance tube. In few colonies workers found to deposit particles of "Oil Mace" on the entrance tube. The length and width of the tube vary according to the strength of the colony and the location of the nest. There is no direct relation between the length of the tube and the age of the colony. The length of the entrance tube varies from 1.2 to 11 cm. The diameter of the entrance tube ranges from 0.7 to 2.4 cm. The wall of the tube is thin and fragile. The entrance tubes are usually directed downwards, but at our surprise one of the entrance tube was found directed upwards. The number of guard bees at the entrance varies according to the strength of the colony.

Entrance tube were absent in some colonies. The number of guard bees at entrance varies from two to sixteen. The shape of the entrance may be slit like, circular, oval or funnel like. Around the entrance tube resin deposit is found in some colonies. This deposit may be in definite patterns like concentric rings or may be irregular. The shape, size and direction of the entrance tube vary according to the nest site conditions. Outer surface of the entrance tube is usually rough and the inner surface is smooth. The recently built apical part of the entrance tube is softer and usually slightly sticky. Usually a single entrance is present at the tip of the entrance tube.

Though rare in some cases more than one entrance is seen on the same entrance tube. In some nests two separate functional entrance tube were observed. Instead of opening directly into the nest the entrance tube opens into the nest through some tunnels called internal tunnel.

4.1.2 Internal tunnel

Internal tunnel are usually branched and are made of soft cerumen. The long internal tunnel helps the stingless bees to effectively defend the colony against intruders.

4.1.3 Resin dumb

The nests of stingless bees are made up of cerumen, a mixture of resin and wax which remains soft for an extended time and is more pliable than beeswax (Wille, 1983; Hepburn and Kurstjens, 1984). In addition to being used for constructing nest forms, cerumen may also be used for repairing nest damage. The resin is collected from resinous plants like Jack fruit, mango tree etc. and wax is secreted by the young worker bees. The chemistry of propolis depends on the diversity of plants from which the bees collect it (Pereira, 2003).

The resin is deposited in the nest at different places especially near the entrance. Guard bees use this resin to prevent invaders (Wille, 1983). In addition to being used for constructing nest forms, cerumen may be taken to make an emergency repair of natural enemy damage (Roubik, 2006). Cerumen is normally made freshly to construct brood cells, involucrum, nest entrance tubes, or storage pots. The wax is taken from a pure wax deposit and mixes it with fresh resin taken from a resin deposit. These may be in several positions near the nest entrance and brood cells; workers mix the materials with their mandibles (Sakagami, 1982). Melipona are keenly interested in returning to a damaged nest and collect resin from resin deposits, and also cerumen and honey, as do many stingless bee genera (Roubik, 2006). Inside the nest resin deposit can be seen at different places especially at the entrance. Three to eight deposits were seen in different nests. Nests with weak wall contain more resin deposit. In pot nest covered with cloth worker bees deposit large quantity of resin.

Table 1. Resin sources for *Trigona iridipennis* Smith in Kerala

Sl. No.	Common Name	Scientific Name	Family	Source of Resin
1	Mango tree	*Mangifera indica*	Caesalpiniaceae	Cut stem
2	Jack fruit tree	*Artocarpus integrifolius*	Moraceae	Cut branches, fruit stalk
3	Tapioca	*Manihot utilissima*	Euphorbiaceae	Petiole
4	Drum stick	*Moringa oleifera*	Moringaceae	Stem
5	Banyan tree	*Ficus bengalensis*	Moraceae	Cut stem
6	Vatta*	*Macaranga peltata*	Euphorbiaceae	Cut stem
7	Kudampuli*	*Garcinia cambogia*	Clusiaceae	Cut stem
8	Cashew	*Anacardium occidentale*	Anacardaceae	Cut stem
9	Sun flower	*Helianthus annus*	Asteraceae	Flower bud
10	Christmas tree	*Arecaria exelsa*	Arecaceae	Cut stem
11	Fig tree	*Ficus roxburgii*	Moraceae	Cut stem

*Name in local language (Malayalam)

4.1.4 Waste Dump

An important feature of the nest architecture of stingless bee is the occurrence of distinct waste deposits (Kolmes and Sommeijer, 1992). In stingless bees resin and waste materials are found heaped on the nest floor (Drummond et al., 1998). In Apis bees the workers defecate outside the nest. In stingless bees latrines are maintained within the nests, where several symbionts live. Adult defecations in the nest are gathered at small latrines; most are consumed by symbiotic organisms (Roubik, 2006). Drainage outlets are maintained in the nests of subterranean stingless bees, such as Meliponula and Plebeiana, and in tree nesting species including Trigona and Tetragona (Sakagami, 1982; Camargo and Roubik, 1991).

Waste is deposited in the nest at different places. They are mainly in the form of granules. These deposits contain dead bees, broken wings, discarded cocoons etc. A maximum of six waste dumps were observed in some colonies. Worker bees carry these waste particles with their mandible and deposit it on the floor of the nest at definite places. Worker bees carry this waste on their mandibles and throw away at distant places at day time.

4.1.5 Food Storage

It consists of pollen pots and honey pots. Honey and pollen pots are oval and same sized (Drumond, 1998). Honey and pollen are stored in separate pots (Figure 3). Usually these pots are intermixed. In some colonies they were found stored separately. Honey and pollen pots are made up of cerumen a mixture of wax secreted by the worker bees and resin collected from plants. They are usually oval in shape, chocolate in colour and are almost similar in size. The cell openings always directed upwards. A large number of pollen pots are usually placed near the entrance and near the brood cluster. The honey and pollen pots are found either separate or intermixed. The height of the pollen pot ranges from 0.7 to 1.2 cm and the diameter from 0.8 to 1.0 cm. The height of the honey pots varies from 0.6 cm to1.1cm and the diameter from 0.7 to 1.2 cm.

Table 2. Internal structure of the nest of *Trigona iridipennis* Smith.

Parameters	Brood cells		Pollen pots	Honey pots
	Worker/drone	queen		
Dimensions				
a) Height	3.72-3.98 mm	7.12-7.43 mm	0.7-1.2 cm	0.6-1.1 cm
b) Diameter	2.92-3.31 mm	4.03-4.27 mm	0.8-1.1 cm	0.7-1.2 cm
Shape	Oval	Oval	Oval	Oval
Colour	Brown to cream	Brown to cream	Dark brown	Dark brown

4.1.6 Brood

The brood area forms the thermal core of a stingless bee nest, where heat may be kept in (or out) by concentric involucra (Roubik, 2006). In cluster type bees, involucrum surrounding the brood is absent (Sommeijer, 1984). The cells and cocoons are cluster type. The advancing front of new cells is roughly horizontal and progress upward through the brood chamber, and all new cells open upwards. In nests having adequate space and a more or less cylindrical cell cluster, the advancing front of cells is somewhat concave (Michener, 1961). In stingless bee nest most of the species arrange brood cells in horizontal combs. In other species the brood cells are positioned in clusters; few species arrange their brood cells in intermediate arrangements (Sommeijer et al., 1982).

The number of brood cells produced per day varies from about 10 in Melipona colonies up to several hundred in *Trigona* species. In *Melipona* species, all bees are born from identical cells, while in most other genera, queen cells are larger than those that produce workers and males (Sakagami, 1982).

Brood cells are spherical or ovoid cells cluster together in group. Cells always open upwards. Egg cups are made of soft cerumen. The newly constructed brood cells are brown in colour which contains egg or larvae. In the pupal stage the worker bees scrap away the cerumen from the brood (Michener, 1974) and the brood looks white or cream in colour. The cerumen scraped from the brood is reused for the brood cell construction. The old cells from which bees had emerged is replaced by new cells (Sommeijer et al., 1984).

The worker and drone eggs look similar. Queens develop in larger cells that are generally, built around the borders of the brood. Queen brood cells differ among the stingless bees. In the genus Melipona, queens are produced in comb cells of the same size and shape as those that give rise to workers and males (Marcia de F.Ribeiro et al., 2003). In Trigona and related genera, queen cells are larger, in length and width, than those of workers and males (Engels and Imperatriz-Fonsaca, 1990). Sometimes the queen cells are found amidst the worker/drone brood. The number of eggs laid per day varies considerably according to the situations of the colony, especially the availability of the food, climatic conditions etc.

Each brood cell is built by several young wax secreting workers. Complete cells have a rim of wax, called a collar, which is used for sealing it after oviposition. The number of cells produced per day varies about 10 in Melipona colonies up to several hundred in Trigona species. In Melipona species, all bees are born from identical cells, while in most other genera, queen cells are larger than those that produce workers and, males (Sakagami, 1982). The height of the worker/drone brood cell is 3.72 to 3.98 mm and the diameter is 2.92-3.31 mm.Height of the queen cells is 7.12-7.43 mm and diameter is 4.03-4.27 mm. Usually one advancing front is present, in few colonies two advancing fronts detected.

4.1.7 Pillars and Connectives

Inside the nest the brood cells and food pots are fixed on definite pillars made up of cerumen. This gives enough space for the worker bees to move between the brood and food pots. The perpendicular structures are called pillars and the horizontal structures are called connectives, which fixes the brood and food pots to the floor and wall of the nest (Wille, 1983). It also attaches brood and food pots among themselves with enough bee spaces between them. The brood cluster is connected to the hive wall by means of pillars which are thicker than those interconnecting the brood cells (Sommeijer et al., 1984).

4.1.8 Nest envelops

If the cavities are extra large stingless bees construct partition walls with special plates called batumen which is made of hard cerumen. The wall of the nest cavity is usually lined with a layer of cerumen called lining batumen (Michener, 1964). A single or multilayered cerumen called involucrum may surround the brood nest. Very rarely an involucrum is seen inside the nest of *Trigona iridipennis*. They usually construct an involucrum if the nest is extra large and also when the lid of the nest slide off. This envelope protects the colony from enemies and also helps to maintain proper temperature and humidity inside the nest.

4.2 Anthropogenic habitats

Trigona iridipennis Smith is seen in forest/wild as well as in natural habitats. The major natural habitats include the living trunks of different tree species especially teak (*Tectona grandis* Linn), jack fruit (*Artocarpus integrifolius*), wild jack fruit (*Artocarpus hirsutus* Lam.), cycas (*cycas sphaerica*) and mango (*Mangifera indica* L) (Table 1). Usually *Trigona iridipennis* Smith occupies in tree trunks of diameter above 30 cm and levels of few centimetres above the ground. They build the nests on trees which have hollow spaces due to decay or rotting of the stem or braches which arise due to insect attack, disease as well as physical damage. In central Travancore majority of the nests in the tree trunks was seen on Maruthu (*Terminalia paniculata*) and at coastal areas it was poovarasu (*Hopea glabra*). It may be because both these trees have lot of cavities in their trunk, as farmers cut their branches regularly for mulching. The nest cavities were cylindrical with lengths up to 90 cm. The nest entrance consist of a tube made up of cerumen, a mixture of wax secreted by young worker bees and resin collected from plants as well as the debris with an average size of 7.0 mm. The inner walls of the nest were lined by black lining batumen and the brood cells were arranged in clusters. For the 43 observations were made for the bees in natural habitats, 13 were on cracks of the stone wall basements of the houses, 11 were on mud plastering of the house walls, 11 were on hollow bricks on walls, 5 on cracks of the external fence while 6 were observed on tree logs. In addition, observations were also made on them in telephone posts where cable box installed, wooden master control box for power supply, air outlet pipes of latrines etc.

Table 3. Trees harboring nests of *Trigona iridipennis* Smith in Kerala.

Sl.No	Common Name	Scientific name	Family
1	Teak	*Tectona grandis* Linn.	Verbanaceae
2	Maruthu*	*Terminalia paniculata*	Combretaceae
3	Poovarasu*	*Hopea glabra*	Dipterocarpaceae
4	Kanjiram*	*Strychnox nuxvomica*	Loganiaceae
5	Neer mathalam*	*Crateva religiosa*	Capparidaceae
6	Jack fruit tree	*Artocarpus integrifolius*	Moraceae
7	Mango	*Mangifera indica* L	Caesalpiniaceae
8	Tamarind	*Tamarindus indica* L	Caesalpiniaceae
9	Manja vaka*	*Albizia lebbeck*	Fabaceae
10	Manjium	*Acacia manjium*	Fabaceae
11	Gul mohr	*Poincinia regia (Delonix regia)*	Caesalpiniaceae
12	Neem	*Melia azedirachta*	Meliaceae
13	Rain tree	*Albizia saman*	Fabaceae
14	Silver oak	*Grevillea robusta*	Proteaceae
15	Red wood	*Adenanthera pavonia*	Mimoceae
16	Coconut	*Cocos nucifera* L	Palmae
17	Arecanut	*Areca catechu* Linn.	Palmae
18	Wild jack fruit tree	*Artocarpus hirsutus* Lam.	Moraceae

* Name in local language (Malayalam)

4.3 Domesticated habitats

The major domesticated habitats for *Trigona iridipennis* Smith across Kerala were wooden boxes (standing as well as hanging), earthen pot and vessel, bamboo nodes, log nests, areca nut stems were inside pith were removed, earthen pipe, PVC pipe, asbestos pipe, and coconut shell. The wooden boxes were made of mostly 'Maruthu' (*Terminalia paniculata*) or any hard wood of comparatively less commercial value. The size and dimensions of the wooden hives were also different. Some boxes have a dimension of 12 x 12 x 34 cm, while the hanging ones were often less dimension (10 x 10 x 36 cm). In some places farmers adopt the compartmentalization of nests, where separate chambers are provided for honey storage as well as brood rearing. Variety of materials were used for hanging the boxes which includes, ropes made out of coconut husk, plastic rope, even insulated electrical wires or steel wire. Usually the ropes were coated with neem oil, grease or used engine oil to repel ants and other walking predators to reach the nest.

The polyvinyl chloride (PVC) pipes used were having a size of 12 cm diameter. The ends of the pipes were closed with end blocks having hole at the centre or coconut shells where the outer surface faces to outside. Holes were usually made on one of the master eye of the coconut shell which is quite easy to pierce and open to get a perfect circular lesion. Earthen pots as well as vessels were employed for keeping *Trigona iridipennis* Smith. Usually pots of different sizes were used depending on the availability ranging from 3-5 litre capacity. The most common one have a neck size of 16 cm diameter and were usually closed using wooden plank. The ring shaped neck was tied with ropes and hanged at the eves of house or sheds.

In olden days men used to collect honey from colonies of stingless bees on trees by cutting down the portion of the timber where the bees were and later axed open. After collecting excess honey the timber was hung down from the eves of house or sheds. Later when there arouse a need for honey, this timber portion would split open and the earlier process of joining together the splinters and hanging it down would be repeated. Today, however, colonies of stingless bees are reared in bamboo nodes, earthen pots, wooden boxes, coconut shells, PVC pipes etc.

4.4 Different types of nests

4.4.1 Bamboo nodes

Two or three node length of bamboo is cut out and then split open through the centre and remove the internodes in between except the internodes at both ends. Colonies of stingless bees with brood, honey and pollen are transferred into the bamboo poles. A small opening is made at one end for the bees to move in and out. Using stingless bee resin all the openings other than entrance can be sealed. Using resin a ring can made and fix around the entrance, which will help the bees in finding the door to the new nest. At sunset when the bees have all entered the nest, the bamboo pole may safely place so that it is protected from direct sun-light and rain (Figure 4).

4.4.2 Earthen pot

At the side of the pot, an entrance is made with the help of a screw- driver for the bees to move in and out. Honey, pollen, brood and bees should be transferred into the pot and then a wooden plank may be used to cover the mouth of the pot. The crevices between the pot and the plank should be sealed up with resin. The pot is to be then hung down in a swing away from the direct sun light or rain.

4.4.3 Wooden Box

The most widely used and most convenient nest for stingless bee is wooden box. Though boxes of different sizes and shapes are in use, it is boxes of 10 cm x 10 cm x 36 cm are ones that is widely used. Any hard wood can be used for the box. All openings except the entrance are to be sealed off with resin. If the boxes are built in such a way that the planks at the bottom and the top are removable; it will be convenient for honey extraction.

4.4.4 Earthen bowl

It is possible to rear stingless bees in earthen bowls having the size of big coconut shells. There are to be two bowls, arranged one above the other. The lower one should be used for depositing the brood and the bees with queen, and the upper bowl bear food pots is to be closed with a wooden plank. All openings are to be sealed with resin. An entrance with a resin lip should be arranged on the lower bowl. The upper bowl should contain an opening having 2 cm diameter for bee passage. As per the availability of honey any number of bowls with a hole at the bottom might be piled up. The new bowls thus added are to be placed in the middle.

4.4.5 Coconut shell

We can rear stingless bees inside big coconut shells that are split at the middle. Such bottom shells is to be used the same way as in the case of earthen bowls where the brood is to be deposited. The upper shell where the eye is opened may now used to cover the bottom shell. This mechanism is to be duplicated with the eye of the shell turned downward, so that the eyes of both the shells touch each other. Inside this shell we may again place pollen, honey etc. and then it is to be closed with other shell of the coconut. All openings are to be sealed off with resin. An entrance hole with a lip is to be arranged on the bottom shell at the centre. During the season of honey flow, the shells of yet another coconut with holes on both ends are to be fitted in the middle of the two sets of shells that were arranged earlier. The holes on the shells are to come in the same line.

4.4.6 PVC pipe

Split a 4 inch PVC pipe through the centre using a cutter and transfer a colony of stingless bee with brood and food pots. Both ends of the pipe can be closed using end caps or coconut shells. Make an opening at one end for bee passage. Seal all the openings except entrance with resin and keep the nest at perfect cool shade.

Table 4. Site description of the selected locations (W1-W21) across Kerala, India for nesting behavioural study during 2011-2013.

Site	Position	Type of hives	Altitude (m)	Rainfall (mm)	Crops
W1 (Ellakkal, Idukki)	10° 0' 10.76" N 77° 3' 59.61" E	Mud wall	741.8	3,446	Pepper, coffee
W2 (Melukavu, Kottayam)	9° 47' 50.96" N 76° 45' 56.69" E	Tree trunk	406.0	3,060	Rubber
W3 (Marayur, Idukki)	10° 15' 49.82" N 77° 9' 57.31" E	Stone wall	965.8	3,446	Ornamental flowers
W4 (Kottathara, Palakkad)	11° 2' 7.29" N 76° 35' 27.38" E	Stone wall	167.9	2,315	Rubber, coconut, areca nut
W5 (Kanjangad, Kasaragode)	12° 18' 47.98" N 75° 5' 45.24" E	Stone wall	21.9	3,517	Rubber
W6 (Sreekandapuram, Kannur)	12° 2' 21.58" N 75° 32' 42.50" E	Stone wall	18.4	3,438	Rubber, coconut
W7 (Vaduvanchal, Wynadu)	11° 32' 58.99" N 76° 13' 44.04" E	Stone wall	876.0	2,481	Pepper, coffee
W8 (Palakkayam, Palakkad)	10° 57' 50.90" N 76° 33' 17.20" E	Stone wall	141.5	2,315	Rubber, coconut, areca nut
W9 (Irinnajalakuda, Trissur)	10° 20' 40.81" N 76° 12' 33.73" E	Stone wall	13.2	2,973	Rubber, coconut
W10 (Kalloorkad, Ernakulam)	9° 58' 10.52" N 76° 41' 1.39" E	Stone wall	42.0	3,422	Ornamental flowers
W11 (Kadammanitta, Pathanamthitta)	9° 18' 40.49" N 76° 46' 21.35" E	Stone wall	113.5	2,572	Rubber
W12 (Chennankari, Alappuzha)	9° 22' 10.19" N 76° 27' 37.22" E	Stone wall	4.9	2,566	Coconut, rice
W13 (Nedumangadu, Thiruvananthapuram)	8° 36' 11.98" N 77° 0' 10.08" E	Stone wall	69.3	1,628	Rubber, coconut
W14 (Karunagapally, Kollam)	9° 3' 15.98" N 76° 32' 7.08" E	Stone wall	10.4	2,310	Vegetables, rubber, coconut
W15 (Mukkom, Kozhikod)	11° 19' 34.57" N 75° 59' 23.49" E	Stone wall	20.9	3,728	Vegetables, rubber
W16 (Karuvarkundu, Mallapuram)	11° 7' 0.01" N 76° 0' 32.03" E	Stone wall	38.1	2,513	Vegetables, rubber
W17 (Nilambor, Mallapuram)	11° 16' 45.94" N 76° 14' 23.20" E	Stone wall	39.3	2,513	Vegetables, rubber
W18 (Mannuthy, Trissur)	10° 31' 44.07" N 76° 15' 44.67" E	Stone wall	23.0	2,973	Vegetables, rubber
W19 (Kuravilangadu, Kottayam)	9° 45' 30.38" N 76° 33' 47.77" E	Stone wall	25.6	3,060	Vegetables, rubber
W20 (Thumbamon, Pathanamthitta)	9° 13' 0.01" N 76° 43' 0.01" E	Stone wall	21.4	2,572	Vegetables, rubber
W21 (Vypin, Ernakulam)	10° 4' 7.31" N 76° 12' 47.33" E	Stone wall	11.5	3,422	Vegetables, rubber

5. Discussion

The Stingless bee nests are great example of colonial life of social insects, which plays a major role protecting from various environmental factors as well as from predators (Pavithra et al., 2012). In addition it provides a microclimate, storage of honey, pollen and rearing young ones. The selection of various natural and adapted anthropogenic habitats may suggest their increasing adaptability and easy rearing in meliponiculture. There is not much recorded evidence about the cultivation history of *Trigona iridipennis* Smith in Kerala, however it may concluded that their culture is popular in rural as well as some tribal areas. The main reason may be the easiness in keeping as well as their high demand in ayurvedic, unani and traditional medicinal preparations.

Workers make wax that is secreted from dorsal glands (Michener, 1974; Sakagami, 1982). Different stingless bee waxes have different chemical properties, and are much simpler than wax of Apis (Blomquist et al., 1985). Cerumen, a product of plant resin mixed with wax and employed exclusively by meliponines, remains soft for an extended time and are more pliable than bees wax that of *Apis mellifera* (Hepburn and Kurstjens, 1984).

Various beekeeping methods preferred by farmers across Kerala for the cultivation of *Trigona iridipennis* Smith. Each nest has its own advantage and disadvantage. During the survey, the most preferred one's were wooden box. Even then according to the easy availability and production cost different nests like earthen pot, bamboo nodes, coconut shell, PVC pipes etc were used. The most preferred natural nesting sites by *Trigona iridipennis* Smith were mud/stone walls in our study. The closeness to the food availability as well as the rigid nature of the habitat may be a reason for the selection of these habitats. The adaptation to various habitats can be governed by environmental factors in addition to the genetic factors (Lima and Dill, 1990; Jaenike and Holt, 1991; Martin, 2001; Cameron et al., 2004; David, 2006). Therefore a better understanding of the genetic nature of *Trigona iridipennis* Smith may able to explain the attractiveness of various habitats.

The nest entrance plays a striking role in the different habitat of *Trigona iridipennis* Smith. The nest entrances were also decorated with animal excreta, fibres, used engine oil, grease, mud particles, paint particles and oil mace particles. In addition with the wax secreted from their bodies, mixed with resins and gums from plants, Greece and used engine oils to make the mould workable and nest thermo resistant.

The nest act as a fortress, regulate temperature and protect from rain where spate spaces were allocated for storing honey, pollen and brood cells.

In earthen bowl and coconut shell methods it is very advantageous to collect honey without causing any disturbance to the brood chamber that is at the bottom. In PVC pipe and coconut shell the colony often desert in summer due to high temperature. Even though the bamboo nodes are alright, they are found to degenerate after few years through the attack of wood borers.

6. Conclusions

Being a social insect, *Trigona iridipennis* Smith shows great diversity in nesting pattern in natural habitats as well as anthropogenic habitats. The shift towards anthropogenic habitats than natural ones may be due to destruction of natural habitats as well as the availability of manmade habitats. Our study also highlights the various cultivation feasibility of *Trigona iridipennis* Smith culture which can be explored more for the demand of stingless bee honey (Dammer honey) in Kerala.

Acknowledgements

The first author is grateful for the cooperation of the management of Mar Augsthinose college for necessary support. Technical assistance from Binoy A Mulanthra is also acknowledged. We also thank all the farmers for who cooperate with us during the study.

References

Amano, K., Nemoto, T. & Heard, T.A. (2000). What are stingless bees, and why and how to use them as crop pollinators?-a review. *Japan Agricultural Research Quarterly, 34*(3), 183-190.

Andualem, B. (2013). Synergistic antimicrobial effect of Tenegn honey (*Trigona iridipennis*) and garlic against standard and clinical pathogenic bacterial isolates. *International Journal of Microbiological Research, 4*(1), 16-22.

Bänziger, H., Pumikong, S. & Srimuang, K.O. (2011). The remarkable nest entrance of tear drinking *Pariotrigona klossi* and other stingless bees nesting in limestone cavities (Hymenoptera: Apidae). *Journal of the Kansas Entomological Society, 84*(1), 22-35.

Basavarajappa, S. (2010). Studies on the impact of anthropogenic interference on wild honeybees in Mysore District, Karnataka, India. *African Journal of Agricultural Research, 5*(4), 298-305.

Biesmeijer J.C., Guirfa M., Koedam D., Potts S.G., Joel D.M. & Dafni A. (2005). Convergent evolution: floral guides, stingless bee nest entrances, and insectivorous pitchers,*Naturwissenschaften, 92*, 444-450.

Boorn, K.L., Khor, Y.Y., Sweetman, E., Tan, F., Heard, T.A. & Hammer, K.A. (2010). Antimicrobial activity of honey from the stingless bee Trigona carbonaria determined by agar diffusion, agar dilution, broth microdilution and time-kill methodology. *Journal of Applied Microbiology*, 108(5), 1534-1543.

Cameron, E.C., Franck, P. & Oldroyd, B.P. (2004). Genetic structure of nest aggregations and drone congregations of the southeast Asian stingless bee *Trigona collina. Molecular Ecology, 13*(8), 2357-2364.

Camrgo J.M.F. & Pedro S.R.M. (1992).Systematics, phylogeny and biogeography of the Meliponinae (Hymenoptera, Apidae): a mini- review. *Apidologie, 23, 509-522.*

Chinh, T.X., Sommeijer, M.J., Boot, W.J. & Michener, C.D. (2005). Nest and colony characteristics of three stingless bee species in Vietnam with the first description of the nest of *Lisotrigona carpenteri* (Hymenoptera: Apidae: Meliponini). *Journal of the Kansas Entomological Society, 78*(4), 363-372.

Choudhari, M.K., Haghniaz, R., Rajwade, J.M. & Paknikar, K.M. (2013). Anticancer activity of Indian stingless bee propolis: an in vitro study. *Evidence-Based Complementary and Alternative Medicine, 2013*(2013), 1-10.

Choudhari, M.K., Punekar, S.A., Ranade, R.V. & Paknikar, K.M. (2012). Antimicrobial activity of stingless bee (*Trigona* sp.) propolis used in the folk medicine of Western Maharashtra, India. *Journal of Ethnopharmacology, 141*(1), 363-367.

Cortopassi-Laurino, M., Imperatriz-Fonseca, V.L., Roubik, D.W., Dollin, A., Heard, T., Aguilar, I., Venturiei, G.C., Eardley, C. & Nogueira-Neto, P. (2006). Global meliponiculture: challenges and opportunities. *Apidologie, 37*(2), 275-292.

Danaraddi, C.S. & Viraktamath, S. (2007). Studies on stingless bee, *Trigona iridipennis* Smith with special reference to foraging behaviour and melissopalynology at Dharwad, Karnataka. *Master of Science Thesis. College of Agricultural Science. Dharwad.*

Danaraddi, C.S., Viraktamath, S., Basavanagoud, K. & Bhat, A.R.S. (2010). Nesting habits and nest structure of stingless bee, *Trigona iridipennis* Smith at Dharwad, Karnataka. *Karnataka Journal of Agricultural Sciences, 22*(2), 310-313.

David, W.R. (2006). Stingless bee nesting biology. *Apidologie, 37*(2), 124-143.

DeMera, J.H. & Angert, E.R. (2004). Comparison of the antimicrobial activity of honey produced by *Tetragonisca angustula* (Meliponinae) and *Apis mellifera* from different phytogeographic regions of Costa Rica. *Apidologie*, 35(4), 411-417.

Devanesan S., Shailaja K.K., Pramila K.S. (2009). Status paper on Stingless bee *Trigona iridipennis* Smith. All India Co-ordinated Research Project on Honey bees and pollinators, Vellayani Centre,Thiruvananthapuram.pp79.

Drumond, M.P., Zucchi R., Yamane, S. & Sakagami S.F. (1998). Oviposition behaviour of the stingless bees XX. *Plebeia (Plebeia) juliani* which forms very small brood batches (Hymenoptera:Apidae,Meliponinae). *Entomological Science*, 1(2), 195-205.

Franck, P., Cameron, E., Good, G., Rasplus, J.Y. & Oldroyd, B.P. (2004). Nest architecture and genetic differentiation in a species complex of Australian stingless bees. *Molecular Ecology, 13*(8), 2317-2331.

Heard, T.A. (1999). The role of stingless bees in crop pollination. *Annual Review of Entomology, 44*(1), 183-206.

Jaenike, J. & Holt, R.D. (1991). Genetic variation for habitat preference: evidence and explanations. *American Naturalist, 137*, S67-S90.

Jose, S. K & Thomas, S. (2012). Stingless beekeeping (Meliponiculture) in Kerala. In: Selected beneficial and harmful insects of Indian subcontinent, (Thomas, K.S

(2012) ed. LAP LAMBERT Academic Publishing GmbH & Co. KG, Saarbruken, Germany.

Kolmes, S.A. & Sommeijer, M.J. (1992). Ergonomics in stingless bees: changes in intranidal behaviour after partial removal of storage pots and honey in *Melipona favosa* (Hym. Apidae, Meliponinae). *Insects Sociaux, 39*(2), 215-232.

Krishnakumar, K.N., Prasada Rao, G.S.L.H.V. & Gopakumar, C.S. (2009). Rainfall trends in twentieth century over Kerala, India. *Atmospheric Environment, 43*(11), 1940-1944.

Kumar, M.S., Singh, A.J.A.R. & Alagumuthu, G. (2012). Traditional beekeeping of stingless bee (*Trigona* sp) by Kani tribes of Western Ghats, Tamil Nadu, India. *Indian Journal of Traditional Knowledge, 11*(2), 342-345.

Lima, S.L. & Dill, L.M. (1990). Behavioural decisions made under the risk of predation: a review and prospectus. Canadian Journal of Zoology, *68*(4), 619-640.

Marisa, H. & Salni, S. (2012). Red wood (*pterocarpus indicus* wild) and bread fruit (*artocarpus communis*) bark sap as attractant of stingless bee (*Trigona* spp). *Malaysian Journal of Fundamental and Applied Sciences, 8*(2), 107-110.

Martin, T.E. (2001). Abiotic vs. biotic influences on habitat selection of coexisting species: climate change impacts?. *Ecology, 82*(1), 175-188.

Michener, C.D. (1974). The social behaviour of the bees. A comparative study. Belknap, Harvard University Press, Cambridge, Mass.404pp.

Mohan, R. & Devanesan, S. (1999). Dammer bees, *Trigona iridipennis* Smith (Apidae: Meliponinae) in Kerala. *Insect Environment, 5*(1), 79-81.

Nair, M.C. & Nair, P.K.K. (2001). Beekeeping by Kanikkars in southern Western Ghats of Kerala. *Indian Bee Journal, 63*(1 & 2), 11-16.

Nair, M.C. (2003). Apiculture resource biodiversity and management in Southern Kerala. PhD. Thesis. Mahatma Ghandhi University, Kottayam, 277.

Pavithra, N., Shankar, R., Jayaprakash. (2012). Nesting pattern preferences of stingless bee, *Trigona iridipennis* Smith (Hymenoptera: Apidae) in Jnanabharathi campus, Karnataka, India. *International Research Journal of Biological Sciences, 2*(2), 44-50.

Raju, A.J.S., Rao, K.S. & Rao, N.G. (2009). Association of Indian stingless bee, *Trigona iridipennis* Smith (Apidae: Meliponinae) with Red-listed *Cycas sphaerica* Roxb. (Cycadaceae). *Current Science, 96*(11), 1435-1436.

Ramanujam, C.G.K., Fatima, K. & Kalpana, T.P. (1993). Nectar and pollen sources for dammer bee (*Trigona iridipennis* Smith) in Hyderabad (India). *Indian Bee Journal, 55*(1/2), 25-28.

Rasmussen, C. (2013). Stingless bees (Hymenoptera: Apidae: Meliponini) of the Indian subcontinent: Diversity, taxonomy and current status of knowledge. *Zootaxa, 3647*(3), 401-428.

Roubik, D.W. (1989). Ecology and natural history of tropical bees. Cambridge university Press, New York.514p.

Roubik, D.W. (2006). Stingless bee nesting biology. *Apidologie, 37*(2).124-143.

Singh, R.P. (2013). Domestication of *Trigona iridipennis* Smith in a newly designed hive. *National Academy Science Letters, 36*(4), 367-371.

Sommeijer, M.J., Beuvens, F.T. & Verbeek, H.J. (1982). Distribution of labour among workers of *Melipona favosa* F: Construction and provision of brood cells. *Insectes Sociaux, 29*(2), 222-237.

Virkar, P.S., Shrotriya, S. & Uniyal, V.P. (2014). Building walkways: observation on nest duplication of stingless bee *Trigona iridipennins* Smith. *Ambient Science, 1*(1), 38-40.

YOUR KNOWLEDGE HAS VALUE

- We will publish your bachelor's and master's thesis, essays and papers

- Your own eBook and book - sold worldwide in all relevant shops

- Earn money with each sale

Upload your text at www.GRIN.com and publish for free